JOHN F. SCHANK • FRANK W. LACROIX
MARK V. ARENA • GORD

T0167767

LEARNING
FROM
EXPERIENCE

—— VOLUME I ——

Lessons from the Submarine Programs of the United States, United Kingdom, and Australia

Prepared for the United States Navy, the United Kingdom's Ministry
of Defence, and Australia's Department of Defence

RAND NATIONAL DEFENSE RESEARCH INSTITUTE

The research described in this report was prepared for the United States Navy, the United Kingdom's Ministry of Defence, and Australia's Department of Defence. The research was conducted within the RAND National Defense Research Institute, a federally funded research and development center sponsored by the Office of the Secretary of Defense, the Joint Staff, the Unified Combatant Commands, the Navy, the Marine Corps, the defense agencies, and the defense Intelligence Community under Contract W74V8H-06-C-0002.

Library of Congress Control Number: 2011939404

ISBN: 978-0-8330-5895-9

Published 2011 by the RAND Corporation
1776 Main Street, P.O. Box 2138, Santa Monica, CA 90407-2138
1200 South Hayes Street, Arlington, VA 22202-5050
4570 Fifth Avenue, Suite 600, Pittsburgh, PA 15213-2665
RAND URL: http://www.rand.org/
To order RAND documents or to obtain additional information, contact
Distribution Services: Telephone: (310) 451-7002;
Fax: (310) 451-6915; Email: order@rand.org

Preface

Large, complex design and construction programs demand personnel with unique skills and capabilities supplemented with practical experiences in their areas of expertise. This is especially true for those designing and constructing conventional and nuclear-powered naval submarines. These vessels require that unique engineer and designer skills be nurtured and sustained and that program managers at all levels be trained and educated so as to create the pool of knowledge and experience to conduct a successful program.[1] In the past, key technical and management personnel in the submarine community were nurtured and sustained through numerous sequential design and acquisition programs supported by growing defense funding. By participating in one or more programs, personnel gained experience to be the leaders in future programs.

But as the operational lives of submarines have lengthened and as defense budgets in most nations have become constrained, new submarine programs are occurring less frequently. In the future, there may be substantial gaps between new program starts, resulting in fewer opportunities for personnel to gain experience managing complex processes and making informed decisions than in the past. Future managers of new programs may not have the benefit of learning from the challenges faced and issues solved in past programs.

Recognizing the importance of past experiences for successful program management, the Program Executive Officer for Submarines from the United States, the Director Submarines of the United King-

---

[1] See Schank et al., 2005a, and Schank et al., 2007.

dom's Defence Equipment and Support organization, and the Director General Submarines from Australia's Department of Defence asked the RAND Corporation to develop a set of lessons learned from previous submarine programs that could help inform future program managers. The research examined the *Ohio, Seawolf,* and *Virginia* programs of the United States; the *Astute* program of the United Kingdom; and the *Collins* program of Australia. This volume summarizes the lessons across those five programs. Other volumes in the series provide descriptions of the specific case studies and the lessons garnered from them:

- MG-1128/2-NAVY, *Learning from Experience, Volume II: Lessons from the U.S. Navy's* Ohio, Seawolf, *and* Virginia *Submarine Programs*
- MG-1128/3-NAVY, *Learning from Experience, Volume III: Lessons from the United Kingdom's* Astute *Submarine Program*
- MG-1128/4-NAVY, *Learning from Experience, Volume IV: Lessons from Australia's* Collins *Submarine Program.*

This research was conducted within the Acquisition and Technology Policy Center of the RAND National Defense Research Institute, a federally funded research and development center sponsored by the Office of the Secretary of Defense, the Joint Staff, the Unified Combatant Commands, the Navy, the Marine Corps, the defense agencies, and the defense Intelligence Community.

For more information on the Acquisition and Technology Policy Center, see http://www.rand.org/nsrd/ndri/centers/atp.html or contact the director (contact information is provided on the web page).

Contents

Summary

Designing and building naval submarines are complex tasks that require organizations with unique skills and expertise. Technical personnel, designers, construction tradesmen, and program managers gain knowledge and experience by working on successive programs during their careers. This will prove difficult in the future as the long operational lives of submarines and the constrained defense budgets of most countries will likely create future gaps in new submarine design and build programs.

Recognizing the importance of past experiences for successful program management, the Program Executive Officer (PEO) for Submarines from the United States, the Director Submarines of the United Kingdom's Defence Equipment and Support organization, and the Director General Submarines from Australia's Department of Defence asked the RAND Corporation to develop a set of lessons learned from previous submarine programs that could help inform future program managers. The research examined the *Ohio*, *Seawolf*, and *Virginia* programs of the United States; the *Astute* program of the United Kingdom (UK), and the *Collins* program of Australia. We developed the lessons from those programs through an extensive review of the appropriate literature in addition to numerous interviews with government and private-sector personnel involved in the programs.

When considering the lessons from the five programs, we must remember that the programs were conducted in different threat and budget environments and with evolving industrial bases for designing and building the submarines. Decisions were made based on the environment at the time, so decisions varied by country and by program.

It is also difficult to judge the success or failure of program decisions. Views change during the conduct of a program and are based on the perspective of individuals. The important point is that the decisions were not necessarily "good" or "bad." Rather, they were or were not fully informed by knowledge of the risks and consequences.

Some lessons are unique to specific programs; others are unique to specific countries. We have tried to identify lessons that apply to all programs and all countries. In some cases, lessons have been identified but not really learned. In other cases, lessons have been learned but forgotten (or ignored). Not only must the United States, the UK, and Australia identify appropriate lessons, they must also heed and remember them. Since cost is typically the metric for judging program success, the majority of the lessons focus on controlling program costs.

Top-Level Strategic Lessons

The top-level strategic lessons apply across all programs and are appropriate for the principal submarine organizations in the government:

- Ensure the stability of the program.
- Be an intelligent and informed partner in the submarine enterprise.
- Establish the roles and responsibilities of the government and private-sector organizations.
- Develop knowledgeable and experienced managers.
- Take a long-term, strategic view of the submarine force and the industrial base.
- Involve all appropriate organizations in any new program.
- Adequately support a new program and make it open and transparent to all.

Lessons When Setting Operational Requirements

Setting the operational requirements will determine the amount of technology risk in a new program. Understanding the cost and schedule impacts of using different technologies or extending operational requirements is important for making informed decisions. Important lessons include the following:

- Remember that the submarine is an integration of various systems.
- Understand the current state of technology to control program risks.
- Involve all appropriate organizations when setting operational requirements.
- Clearly state operational requirements as a set of performance goals.
- Determine how to test for the achievement of desired operational requirements.

Lessons When Establishing an Acquisition and Contracting Environment

An interactive relationship between the government and the prime contractor is necessary for the success of a program. Key lessons for establishing an effective acquisition and contracting environment include the following:

- Consider a single design/build contract for the first-of-class.
- Use a contract structure with provisions for handling program risks and incentivizing the contractor to achieve cost, schedule, and performance goals.
- Develop realistic cost and schedule estimates.
- Decide on government-furnished equipment.
- Develop a timely decisionmaking process to manage change.
- Establish an agreed-upon tracking mechanism and payment schedule.
- Include an adequate contingency pool.

Lessons When Designing and Building the Submarine

To some degree, lessons for the design/build process overlap the lessons that emerged from the programs' earlier stages. These lessons include the following:

- Involve builders, maintainers, operators, and the technical community in the design process.
- Complete the majority of the design drawings before the start of construction.
- Ensure sufficient oversight at the design and build organization.
- Specify and manage adequate design margins.
- Develop an integrated master plan for design and build.
- Track progress during the design/build process.
- Design for removal and replacement of equipment.
- Conduct a thorough and adequate test program.

Lessons When Establishing an Integrated Logistics Support Plan

Operating and supporting new submarines after they enter service account for the vast majority of their total ownership costs. Therefore, it is imperative to establish an integrated logistics support (ILS) plan for the new submarines.

- Establish a strategic plan for ILS during the design phase.
- Consider ILS from a navy-wide perspective rather than a program-specific vantage point.
- Establish a planning yard function and a maintenance and reliability database.
- Plan for crew training and transition to the fleet.
- Maintain adequate funding to develop and execute the ILS plan.

Acknowledgments

This research was requested by the submarine organizations of the United States, United Kingdom, and Australia. For the United States analysis, we appreciate the support and guidance of RADM William Hilarides, the PEO for Submarines when the research was initiated, and RADM David Johnson, the current PEO for Submarines. We also thank Ann Birney, Director, International Programs, who identified key individuals to interview and facilitated those meetings. VADM (Ret) Paul Sullivan, RADM (Ret) Millard Firebaugh, and program managers from Electric Boat provided helpful comments on the draft of the U.S. case studies.

For the United Kingdom analysis, we thank RADM Simon Lister, Director Submarines, and Jonathan Swift, Head Submarine Production in the United Kingdom, for their guidance and support. Mark Hyde identified key individuals to interview and, with Millard Laney, facilitated those meetings. Tony Burbridge, Jonathan Swift, Mark Hyde, and Muir Macdonald provided many constructive comments on the draft of the UK case studies.

For the Australian analysis, we thank RADM (Ret) Boyd Robinson while he was Head, Maritime Systems Division, of Australia's Defence Materiel Organisation. We greatly appreciate his guidance during the study. We also are appreciative of the efforts of CDRE Rick Longbottom, previous Director General Submarines, and Bronko Ogrizek, current Director General Submarines, for their assistance and support. Special thanks go to Ray Duggan, who took on the job of organizing the interviews. Peter Yule and Derek Woolner, authors of the informative history of the *Collins* program, graciously shared insights they gained while gathering background information for their book.

We also thank the numerous people we interviewed from the three countries who shared their time and experiences with us. They include government and private-sector personnel who are or were important managers or observers of the submarine programs we examined. At RAND, Deborah Peetz provided support in identifying and obtaining reports and background information on the various submarine programs. Paul DeLuca of RAND and RADM (Ret) Phil Davis provided technical reviews of an earlier draft of the report and offered several constructive comments that greatly improved the report.

Of course, any errors of omission or commission in the document are the sole responsibility of the authors.

Abbreviations

ASC   Australian Submarine Corporation (now ASC Pty Ltd.)

BCWP   budgeted cost of work performed

BCWS   budgeted cost of work scheduled

CFE   contractor-furnished equipment

COATS   command and control system module off-hull assembly and test site

C4I   command, control, communications, computing, and intelligence

EVM   earned value management

GFE   government-furnished equipment

GFI   government-furnished information

ILS   integrated logistics support

MOD   Ministry of Defence

NSRP   nuclear steam raising plant

OEM   original equipment manufacturer

PEO   Program Executive Officer

RAN   Royal Australian Navy

SSBN   nuclear ballistic missile submarine

SSGN   nuclear cruise missile submarine

SSN   nuclear attack submarine

UK   United Kingdom

CHAPTER ONE

Introduction

Lessons from past experiences are an important tool in preparing managers to successfully lead future programs. This is especially true for managing complex military programs governed by rules, regulations, procedures, and relationships not typically found in commercial projects. In the past, new programs started frequently, giving junior-level managers the opportunity to gain experience and preparing them for more senior management roles in future programs. However, because current naval platforms now have longer operational lives and defense budgets are more constrained, the time between new program starts has lengthened. Managers of new programs often do not have the benefits of experience gained on previous programs. In this environment, it is important that lessons, both good and bad, from previous programs be captured and provided to future program managers, senior navy[1] decisionmakers, and technical resource managers.

Recognizing the need to document lessons from past programs to provide insights for future program managers, the submarine organizations of the United States, the United Kingdom (UK), and Australia asked the RAND Corporation to codify lessons learned from past submarine design and acquisition programs. The research examined the *Ohio, Seawolf,* and *Virginia* programs of the United States; the *Astute* program of the United Kingdom; and the *Collins*

---

[1] Throughout this document, we capitalize the word "Navy" whenever we refer to the U.S. Navy or to a specific country's naval force (such as the Royal Navy). We lowercase the word when we are referring generically to several countries' naval forces or to naval forces in general.

program of Australia. This volume summarizes the lessons across those five programs.[2] In it, we develop lessons identified both in previous reports on the various programs and in numerous interviews that RAND conducted with past submarine program managers and submarine personnel at military and government organizations, as well as at private-sector organizations that design and build submarines in the three countries. We were particularly interested in

- how political, budget, and operational environments influenced decisions made during the program
- how operational requirements guided the design and related to the technologies available at the time
- the contracting and acquisition processes used during the program
- how the private-sector industrial base that designs, builds, and maintains submarines and their systems changed over time
- the interactions between the government and naval organizations and the shipbuilding industrial base
- how integrated logistics support (ILS) plans were developed during the design and construction of the submarines to support the new submarines after they entered service
- how other issues, both internal to the program and external, influenced decisions and outcomes.

The lessons we strive to identify are managerial in nature, not technical. For example, we do not focus on why a specific valve or pump was chosen but rather on how the program was managed, the issues that affected management decisions, and the outcome of those decisions.

Country Differences

The three countries have different histories in new submarine programs for their naval forces. The United States and the UK have designed

---

[2]   See Schank et al., 2011a, 2011b, and 2011c.

and built numerous classes of conventional and nuclear powered submarines over the last century. The United States currently operates the *Los Angeles*, *Seawolf*, and *Virginia* classes of nuclear attack submarines (SSNs) and the *Ohio* class of nuclear ballistic missile submarines (SSBNs).[3] *Virginia*-class submarines are currently being built, and the United States is starting the design effort for a new submarine to replace the *Ohio*-class SSBNs (named the Ohio Replacement Program). Key for the United States was the continuous stream of new programs. The UK operates the *Trafalgar* and most recently *Astute*-class SSNs and *Vanguard*-class SSBNs. The UK is currently building additional submarines in the *Astute* class and is starting the design effort to replace the *Vanguard*-class SSBNs (named the *Successor* program). As we will describe, there was a substantial gap between the design and build of the *Vanguard* class and the start of building the *Astute* submarines. After operating submarines built in the UK for several decades, Australia built the *Collins*. Based on a Swedish design, it is the first class of submarines built in Australia, and the Royal Australian Navy (RAN) operates six of the conventionally powered vessels. Australia is starting the process of defining a new class of submarines (named the SEA 1000 program) to replace the *Collins* class. Defense leaders there have not yet determined whether the SEA 1000 will be an Australian design or be based on the design of another country.

The industrial base that designs and builds submarines is also different in each country. General Dynamics Electric Boat and Huntington Ingalls Industries–Newport News Shipbuilding[4] have designed and built U.S. submarines for several decades. The two companies, once competitors, now share in the build of the *Virginia* class and will most likely team for at least the design of the new SSBNs. The UK shipyard at Barrow, now owned by BAE Systems, has built the vast majority of UK submarines. However, before *Astute*, the initial phases of design were accomplished by government and naval organizations,

---

[3] The United States also operates four *Ohio*-class submarines that were converted from SSBNs to nuclear cruise missile submarines (SSGNs).

[4] Northrop Grumman spun off its shipbuilding sector in March 2011, forming the new Huntington Ingalls Industries.

and the shipyard then turned the contract designs into the detailed design products needed for construction. That process changed for *Astute*, with the prime contractor becoming responsible for all phases of the design effort. Also, the ownership of the Barrow shipyard changed hands twice during the *Astute* program. Australia established the Australian Submarine Corporation (ASC) to design, build, and provide in-service support for the *Collins* class and developed a greenfield site for the construction and deep maintenance of the submarines. It is yet unclear what role ASC will play in the SEA 1000 program.

Differing Threat Environments

The five programs we examined were conducted in differing environments. The *Ohio* program faced the Soviet threats of the Cold War. The *Seawolf, Astute,* and *Collins* programs started during the Cold War but saw their end before the submarines were built. The *Virginia* is a post–Cold War program. The budget environment has also changed over the past 30 years. The large submarine force structures of the United States and the UK have been dramatically reduced, and the funding available for defense projects has become constrained. Finally, the industrial base for designing and building submarines has significantly changed in the United States and the UK. The design and construction industrial base created in Australia to build *Collins* has now evolved into a maintenance and repair industrial base.

Decisions on the required operational capability and the conduct of a new submarine program are made in the context of the environment at the time. Therefore, decisions viewed as appropriate for one program may appear inappropriate for another. Also, decisions vary by country. One country may choose a certain path for a new program for any number of reasons; another country, for equally valid reasons, may choose a different path. The important point is that decisions made during the conduct of a new program are not necessarily good or bad; however, decisionmakers may not have been adequately informed or may have underestimated the impact of various factors.

Other Considerations

For many reasons, it is often very difficult to judge the success of a specific program. Success can be measured in performance, cost, or schedule terms; one person's view of success can differ greatly from the views of others. Also, early "failures" can turn out to be "successes." For example, all five programs had tenuous beginnings with cost and schedule problems. Yet some, such as the *Virginia*, are viewed as successful. It is even more difficult to identify the specific actions or decisions that contributed to success or non-success; in any program, many factors interact. We had to keep this interaction in mind as we sorted through the lessons of the programs, and future program managers should also bear it in mind as they face the challenges of a new program.

Some lessons are specific to programs; others are specific to countries. We try to define lessons that apply to all programs and countries, noting where the lesson, or the decisions based on the lesson, may vary for different programs or different countries. Also, although the lessons are based on submarine programs, we believe they are equally valuable to leaders and managers of future surface ship programs.

In this document, we distinguish between government and private-sector organizations. Government includes the military and civilian organizations that make decisions on managing and funding submarine programs. In the United States, the U.S. Navy manages a program and signs contracts with the private sector to design and build submarines. The U.S. Navy must interact with civilian organizations within the Office of the Secretary of Defense and the Congress. Program managers are typically naval officers. In the UK, the Defence Equipment and Support organization (referred to as Customer One) buys submarines and provides them to the Royal Navy (referred to as Customer Two). Programs are typically managed by civilians. In Australia, the Ministry of Defence contracts with the private sector and provides new submarines to the Royal Australian Navy. Programs are managed by either military or civilian personnel. Private-sector organizations include the firms that design and build the submarines as well as the various vendors that provide services and products to a submarine program.

Controlling Program Cost

Program managers, both in the government and the private sector, strive for successful programs. Success is often measured using cost, schedule, and performance metrics; the acquisition cost of the program is typically the primary focus. The majority of the lessons from the five programs focus on controlling program costs. Cost management must occur throughout a program, from the early stages of concept development through the operational life of the submarine.

Early estimates of program cost are always preliminary and typically have a large degree of uncertainty. But these early cost estimates tend to "stick" with a program throughout its life. Therefore, both the government and the private sector should develop early estimates of program costs that are informed by all relevant historical data and that account for various program risks. Early cost estimates should also consider the status of the industrial base and how well it is positioned to meet program demands.

The desired operational requirements for the new submarine directly affect design and construction costs. An increase in desired performance typically introduces more technical risk in a program, leading to increased costs. The government must understand the relationships between desired performance and cost and set goals that should keep the program within cost constraints. The government should also use the contracting structure to incentivize private-sector contractors to design and build the submarine in the most cost-effective manner.

Controlling costs is most important during the design/build process. An integrated master schedule should define the key milestones and tasks from initial design to delivery of the platform. The start of construction should be keyed to the completion of a large percentage of the design drawings. Therefore, it is important to tightly control any changes to the design products once they are developed. But it is often better to extend the schedule by delaying the start of construction than to start the build of the submarine before the design is set and risk future cost increases due to necessary rework.

The acquisition cost of the platform is typically the metric by which programs are judged. However, the initial acquisition cost is usually far less than the operating and support costs over the life of the submarine. Program managers must weigh the trade-offs between initial design and construction costs and the total ownership costs over the life of the program.

We do not intend to suggest what a new submarine design and construction program should cost; program costs are based on many factors. We do, however, hope that the lessons set forth here can inform future program managers on how best to control costs.

Organization of This Document

To be useful, lessons should be categorized along different dimensions, although many of the lessons run through whatever categorization is used. Chapter Two summarizes lessons that are top-level or strategic in nature versus program-specific. Chapter Three provides lessons that are appropriate for setting the requirements for a new submarine. Chapter Four provides lessons for establishing the acquisition environment and contracting details of a new program. Chapter Five summarizes lessons that apply during the design and build of the new submarine, and Chapter Six addresses lessons for planning integrated logistics support. Chapter Seven summarizes the key lessons for future submarine program managers.

CHAPTER TWO

Top-Level Strategic Lessons

The lessons in this chapter are appropriate for the top-level management of all submarine programs. They are applicable to the U.S. Navy, the UK Defence Equipment and Support organization, and the Australian Department of Defence. These top-level lessons go beyond a single program or a single point in time and stress the long-term view of a nation's overall submarine enterprise.

Ensure the Stability of the Program

One overarching lesson from the various programs is the importance of program stability. It is key to a program's success. Stability applies in many areas—consistent funding, a long-term build strategy, fixed operational requirements, stable and capable program management, and an integrated partnership between the navy and the shipbuilders. Program stability is not sufficient for program success, but it certainly is a necessary attribute that greatly contributes to the success of a program.

Be an Intelligent and Informed Partner in the Submarine Enterprise

The most important lesson from a strategic perspective is that the government, like consumers of all products, must be intelligent and informed in its dealings with the private-sector organizations for the design, build, and support of naval submarines. We avoid the term "customer" with this lesson, since the countries now recognize that they are partners with the private sector rather than adversaries.

This lesson is most appropriate for the UK and Australia, although it should not be forgotten by the U.S. Navy. The UK's Ministry of Defence (MOD), which, before the *Astute* program had maintained significant technical resources to provide knowledge, expertise, and oversight for new submarine design and build programs, lost much of that capability when it dramatically reduced or eliminated government organizations and transferred responsibilities to the private sector. Its "eyes on, hands off" philosophy, partially motivated by the contracting environment, did not work as planned and left it blind to the problems being faced by the prime contractor in developing the design and building the submarines. It has since recognized the fallacy of not being an informed partner in the *Astute* program and has started to rebuild lost capabilities.

Inasmuch as the *Collins* represented Australia's first domestically built submarine, Australia neither fully understood nor was fully prepared for the responsibilities of being a parent navy.[1] Previously, the UK had designed and built the submarines operated by the RAN and provided the majority of the maintenance planning and spare parts for those submarines, but the *Collins* presented a new environment in which Australia was the only operator of the vessel. For the first time, the government had to assume responsibility both for the build of the submarines and their logistics support once they entered service. In the future, Australia must continue to build its technical and oversight resources to successfully manage new submarine programs and the support of the *Collins* class.

The U.S. Navy has typically been a technically knowledgeable and experienced organization when managing submarine programs, and it has used the combined resources of its various commands and laboratories to guide and inform those programs. But when the Cold

---

[1] A *parent navy* operates and supports a submarine that was largely designed and built in-country. Typically, that country is the only country that has that submarine in its force structure.

War ended, funding constraints resulted in a reduction in personnel resources at the Naval Sea System Command and the various submarine centers of excellence. This reduction in technical resources presented challenges for the U.S. Navy in continuing its role as an informed customer. It is now rebuilding much of this lost capability.

For the government to be an intelligent partner, its organizations need the support of experienced technical personnel. Both the civilian and military sides of government should fund centers of knowledge and expertise in areas such as hull dynamics, propulsion systems, signatures, combat and communications systems, and safety of operations. Many, if not all, of these knowledge centers should be in the government; however, academia and the private sector can augment or substitute in some technical areas.

To be an informed partner, the government must also understand past and current costs for the design and build of submarines and be able to adequately estimate the cost of future submarine design and build options. The government needs to understand what factors drive costs and how different technical or managerial decisions can affect those costs. Collecting and organizing historical cost data, using the data to project future costs, and developing internal cost estimating capabilities should be improved in the three countries to varying degrees.

Maintaining adequate technical resources will be challenging for all three countries. Defense budgets will continue to face pressure, and gaps may exist between new submarine program starts. Each country must determine how best to meet those challenges but must also remember that some level of technical support funding is needed to manage the maintenance and modernization of in-service submarines. Being an informed customer also requires some level of pure technology funding when new systems and processes are explored. As allies, the three countries should also discuss how best to integrate and share their technical resources.

In addition to the aspects discussed above, the following lessons will also help the government be an intelligent and informed partner in the submarine enterprise.

Establish the Roles and Responsibilities of the Government and Private-Sector Organizations

The various roles and responsibilities in a new program basically come down to who should assume any risks that may arise. Assuming risks should result in decisionmaking authority. The specific roles and responsibilities of the government and private sector and the locus of the final decision authority in various areas must be firmly established at the start of a new program. The responsibility for different risks should remain constant from program to program so that all organizations clearly understand how a new program will be conducted. However, there may be circumstances that suggest moving responsibility for certain risks from the government to the private sector. Any such changes should be informed by a thorough analysis of how they might alter responsibilities and by a clear plan for the transition. Changes need to be adequately funded and the entities performing the activities need to be fully qualified to implement the changes.

The U.S. Navy has established clear roles and responsibilities with the private sector, which have changed slightly over time. Prior to the *Ohio* program, the U.S. Navy played the primary role in developing conceptual and preliminary designs. Due to the drawdown in the U.S. Navy's technical resources, the private sector now plays the main role in developing the early designs for a new submarine.

Australia had to address this issue for the first time in the *Collins* program. As would be expected with a complex program undertaken for the first time, it proved difficult to determine which party had responsibility for certain risks and where final decisionmaking authority on design and build issues should reside. With the *Collins* experience as a guide, Australia seems better positioned to establish proper roles and responsibilities for the SEA 1000 program.

The UK experienced a major change in the responsibility for certain risks at the beginning of the *Astute* program. Due to pressure to reduce the government's costs for running a new program and the belief that the private sector could accomplish certain tasks at lower cost, many responsibilities previously held by government were transferred to the private sector. For example, the role of design authority,[2] which had been filled by the MOD in previous programs, was assigned to the prime contractor. This was not necessarily a bad decision. The problem was that the private sector was ill prepared to assume this new role, and the two sides did not develop a plan for the transfer of responsibilities. Also, the MOD assumed that it could stand back, let the prime contractor make key decisions on options, and pick up a submarine that met the requirements when the prime contractor had finished construction and testing. It adopted an "eyes on, hands off" policy (although with the drawdown of oversight resources at the shipyard, the MOD effectively lost an "eyes on" capability).[3] The MOD has recognized the problems caused by transferring design authority to the prime contractor, and it will assume the design authority role when the first three submarines enter service and for the build of the remaining submarines in the class.

The key issue with assuming responsibilities is being proactive in managing risks. The government must identify where risks exist and develop a plan to mitigate those risks. And risks must be managed throughout the program—from the initial setting of requirements through the design and build of the submarine, to the acceptance of the submarine by the government.

---

[2] There are various "authorities" in a new program. For example, the U.S. Navy distinguishes between design authority and technical authority. The *design authority's* role is to forward to the designer the design specifications or rules. These are usually based upon the submarine concept selected from the concept studies that preceded the design effort. The design authority must be consulted and approve any proposed changes to the design specifications. In contrast, the *technical authority* is the subject matter expert in various areas, such as the submarine hull, mechanical and electrical engineering, submarine safety, and ship design and engineering. The technical authority is responsible for establishing technical standards in each area and evaluating the risk if a design does not conform with technical standards during design and construction. To be effective, the design and technical authority roles require skilled and experienced staff who have predominantly technical and engineering expertise.

[3] The fixed-price contract for the *Astute* program also tied the hands of the MOD. The MOD was reluctant to impose conditions or mandate changes to the design of the submarine for fear it would lead to cost increases.

Certain risks remain the sole responsibility of the government. These include obtaining the desired military performance from the new submarine and ensuring safety of operations. The government should also strive to deliver the overall program on time and within budget. It shares this risk with the prime contractor and must use all available tools to monitor contract performance, interact with the contractor, and optimally incentivize the builder to meet schedule and cost milestones.[4]

Because it always will need to shoulder certain risks, the government should assume the following responsibilities:

- Set the operational requirements for the new submarine by working with industry, the navy, and other stakeholders.
- Assess safety and technical issues in accordance with the government's policy on safety risks.
- Oversee and monitor the design process to ensure requirements and standards are met and, when necessary, provide concessions to those requirements.
- Oversee and monitor the build process to ensure that the submarines are delivered on schedule and at projected cost.
- Ensure submarine construction quality and acceptability by developing a testing, commissioning, and acceptance process so that the submarines are delivered to the contract specifications and requirements.
- Ensure through-life submarine safety and maintenance and post-delivery control of the design and construction of the submarines in the class.
- Ensure that the model for logistics support fits the country's current and projected infrastructure for maintaining its ships and submarines.

---

[4] The prime contractor also faces risks if it does not efficiently deliver a cost-effective submarine; however, while the prime contractor may go out of business, the government is still responsible for the defense of the nation. Also, there are risks to the prime contractor if the submarine is unsafe, but the government is ultimately responsible for the health and well-being of the sailors.

As an example, design authority responsibility should most likely remain with the government. The acceptance of the ship as safe for operations is the responsibility of the government. It must, therefore, assume the risks associated with the design authority process.

Overall, the government and the private sector must establish an interactive partnership in which information and issues flow freely. Effective interactions will help the government better understand the product it will receive and help the prime contractor develop a product that better fits the navy's needs. In some cases, the government must assume risks; in other cases, the prime contractor should assume them; finally, in many cases, risks should be shared between the government and the prime contractor.

Develop Knowledgeable and Experienced Managers

Successful programs have experienced and knowledgeable people in key management, oversight, and technical positions. Growing future program managers and technical personnel within the civilian and military branches of government requires planning and implementation far in advance of any one specific program. Promising officers, especially engineering duty officers, and civilian personnel must be identified early in their careers and provided suitable education and assignments to ongoing programs at a junior management level. Assigning people who have "earned their stripes" on one program is critical to the success of the next program.

Just as knowledgeable and experienced people are needed in the government, so experience and knowledge are also needed at the prime contractor and major subcontractors. The rapid change in ownership of the company that designed and built the vast majority of the Royal Navy's nuclear submarines, coupled with the movement of key people from the submarine sector to other career fields, resulted in a lack of experienced managers and technicians at the prime contractor for the *Astute* program. The government should encourage, and possibly require, that the prime contractor grow and maintain experienced managers and technical leaders.

Another important aspect is continuity in leadership and in team composition. Managers, leaders, and team members in the government and the industrial base should stay in a program long enough to gain knowledge of the program and maintain its goals.[5] Frequent changes in leadership, which occurred in both the *Astute*'s prime contractor and the *Collins'* prime contractor and government program office, can degrade a program by introducing managers with different goals and strategies from those of their predecessors. Although personnel changes are inevitable, especially for military personnel, they should be minimized to the extent possible, and when new government or private-sector leaders are brought into a program, they should possess knowledge and experience similar to that of the individuals they replace.

Providing early experiences for future program managers is a challenge for Australia inasmuch as the *Collins* program has been the only opportunity for civilians and military personnel to gain expertise. Many of the government personnel involved with the *Collins* have retired or are reaching the end of their careers. It will be important for personnel who were involved with the *Collins* program to be assigned to the SEA 1000 program. Australia may also require assistance from allied countries and their submarine design and build organizations when the RAN's new submarine program begins.

The UK MOD has a policy of growing experienced "generalists" to manage future programs.[6] Although this may work for certain management positions where experience in a management area is the key requirement, technical "specialists" must also be grown and should gain their expertise through working on multiple submarine programs.

Growing knowledgeable and experienced people will be a challenge in the future: Budget constraints may result in fewer new programs on which young officers can gain experience, and force structure reductions may lead to a smaller pool of submarine officers. It is

---

[5]   The UK's Smart Acquisition initiative suggested a minimum of four years for program assignments.

[6]   For example, an Army brigadier general was the first program manager for the Type 45 air warfare destroyer program and a civilian with experience in commercial shipbuilding was one of the early program managers for the UK's *Queen Elizabeth*–class aircraft carrier program.

therefore important that the civilian and military branches of government identify the most promising young civilians and junior officers for future management positions and provide learning experiences for them.

Take a Long-Term, Strategic View of the Submarine Force and the Industrial Base

A specific program is only one step in developing a successful military capability and sustaining the industrial base capacity that provides and supports that capability. The government must take a long-term view and understand how a specific program nurtures and feeds the overall strategic plan for the submarine force.

A new submarine does not remain static once it is delivered to the force. Technologies change, new capabilities are needed, and new threats emerge and evolve. These evolutions require experienced designers and engineers to maintain a technology/capability edge and to update existing platforms with new technologies and new capabilities. In the United States, the improved *Los Angeles* class, the conversion of the *Ohio*-class SSBNs to SSGNs, and the construction of the USS *Jimmy Carter* are examples of how original designs were modified for new missions and capabilities. At some point, new classes of submarines must be designed and constructed.

Both the technical community—the civilian and military engineering directorates and laboratories, test centers, and centers of excellence that support submarines—and the industrial base that designs, builds, and maintains submarines must be sustained at some level so they can provide the required capabilities when needed. This is particularly important for the submarine industrial base because submarine design and construction requires specific skills that cannot be sustained by surface ship programs. Design/build personnel and facilities in the private sector also must be sustained so they can support future submarine design efforts.

Australia developed a submarine construction capability with the creation of the ASC. But the country had no plans on how to sustain that capability once the *Collins* boats were built. In the UK, the substantial gap between design and build of the *Vanguard* class and the start of the *Astute* program was a big contributor to the problems faced by the *Astute* program. This led to a situation in which submarine design and build skills atrophied in the UK, resulting in a costlier and lengthier *Astute* procurement effort. The issue is not that the gap should have been avoided but that the MOD neither anticipated the impact of the gap nor factored the need to rebuild its industrial base capability into the cost and schedule estimates.

In the future, there are likely to be similar gaps due to constrained defense budgets and the long operational lives of submarines. Governments must decide at what level to sustain sufficient resources and expertise during those gaps to allow reconstitution when needed. There are costs and benefits of sustaining various levels of skilled and experienced resources.[7] In addressing these options, governments must be prepared to estimate the implications of a gap on future programs as well as the cost of sustaining resources during a gap.

Sustaining submarine design, build, and technical support resources will be a challenge in the future, but it is a challenge that governments must face. Funding and supporting concept studies for evolutions of existing platforms or for developing new classes of submarines is needed to sustain and nurture these key design resources. These efforts should go beyond the shipbuilders to include major vendors that support submarine design and construction. The history of past programs reinforces the need to maintain a healthy supply base, especially in the submarine community, where many skills are unique and cannot be supported by surface ship programs. Some form of collaboration among the three countries may be a viable way to sustain design and construction skills that could be available when needed by any of the countries.

---

[7] See Schank et al., 2005a, 2005b, and 2007.

Involve All Appropriate Organizations in a New Program

The program and the procurement agency must be supported by adequate technical, operational, and management expertise. The program must have people from the fleet with experience in submarine operations and maintenance, from the research and technical community knowledgeable in the areas of hull, mechanical, and electrical systems as well as propulsion, signature, and survivability issues and from the construction shipyard(s) that understand the potential problems with building certain aspects of a design to identify risks and solutions early and throughout the program. The program should plan on spending the time necessary to ensure that the program's philosophy and underlying principles (e.g., cost control and low technical risk) are clear to all participants and emplaced at all levels. In addition, the program manager should be empowered with appropriate decisionmaking authority (e.g., for change control).

One criticism of the *Collins* program was the absence of the technical community early on. Similarly, the UK's *Astute* program did not involve operators, builders, or maintainers to an appropriate degree during the early stages of the program. Some of the problems with these programs may have been alleviated if they had used a design/build philosophy—involving operators, maintainers, builders, and key suppliers—during the detailed design stages. Early involvement of builders, as well as operators and maintainers, not only helps identify requirements up front but also flags potential problems and their possible solutions.

Another problem often mentioned during our interviews was the lack of integration of the appropriate organizations during a program. Getting the right organizations and personnel involved entails co-locating people from the scientific community, the designer, the builder, the operators, and the maintainers. This fosters engagement and teamwork among all parties. That said, a strong program management structure is needed to oversee and adjudicate the interests of the different groups. Modern communications help bring people together from various locations, but face-to-face interactions are often necessary for effective decisionmaking.

The appropriate people and organizations in the U.S. Congress, the UK Ministry, and the Australian Parliament should also be informed of programmatic decisions and the status of a program. This is the focus of our next lesson.

Adequately Support a New Program and Make It Open and Transparent to All

A new submarine program needs a range of supporters both outside the program and inside government and the submarine community. Political support is most important for the advancement of a new acquisition program. Without the support of the politicians, sufficient funding may not be available to adequately conduct the program. Support must also come from members of the scientific community that possess the technical knowledge needed to make informed decisions and from the public. One lesson from the *Collins* program is the need to effectively manage the media; the bad press that accompanied the *Collins* effort still taints the program in the mind of the general public. Finally, support must come from within the navy. The RAN was not adequately supportive during the early stages of the *Collins* program.

Full disclosure during the program is necessary to obtain government, industry, and public support. There should be periodic feedback to government decisionmakers and to the public on how the program is progressing. Such feedback is especially important when there are unanticipated problems. In this regard, a good media management program is necessary. Effective communications with the press, academia, and government must be proactive, not reactive. Program managers must proactively ensure that all parties are well informed in advance of positive and negative developments and their associated implications.

CHAPTER THREE

Lessons When Setting Operational Requirements

New submarine program managers should seek to reduce risks to the maximum extent possible. In this regard, an important aspect of a new program involves decisions made early in the program about the desired operational performance of the new submarine. These early decisions influence the degree of technology risk for the program and can influence the likelihood of a program's success or failure. Pushing technology frontiers in too many areas will make it more risky to meet program cost and schedule goals.

With respect to technology, the United States and the UK typically have adopted an evolutionary strategy on new programs rather than a revolutionary approach that pushes multiple technology areas. The majority of new submarine classes in both countries have used the best systems available at the time to progressively improve the performance of existing classes. Often, one new technology area was included in a new class, thus reducing technology risks. There have, of course, been exceptions, which have often led to problems.

The U.S. *Seawolf* program and the early stages of the UK *Astute* program (when it was known as SSN20) attempted to make significant gains in operational performance (higher speeds, lower signatures, greater diving depths, and increased payloads) in the face of increased Soviet capabilities during the Cold War. The *Astute* program scaled back the ambitious gains in performance when the Cold War ended and returned to an evolutionary approach that utilized various systems from the existing UK classes of submarines (with the SSN20 being renamed Batch 2 *Trafalgar* class). However, the UK MOD and the prime contractor greatly underestimated the effort involved in integrating systems from various existing submarines. The United States went forward with the *Seawolf* program, which was plagued by cost and schedule problems resulting from having to develop new technologies to meet the desired gains in operational performance. The program was truncated with only three submarines built.

Although Australia had no experience in submarine design and build prior to the *Collins*, that program also attempted to significantly push technology. *Collins* was based on a Swedish design, whose size had to be stretched to accommodate the platform's desired operational requirements. Although small increases in size or other physical capabilities are typically achievable, significant increases result in major design changes. Also, the combat system requirements for *Collins* were well beyond the computing state of the art at the time. The combat system was a major problem faced by the *Collins* program and was overcome only after the United States shared its combat system technology with the Australians. On the positive side in the *Collins* program was the development of a successful state-of-the-art ship control system.

Prototyping is a second method programs have used to better understand new technologies and how they could be incorporated into a new submarine platform. At one time, programs would build large wooden mock-ups of the new submarine to check for clearances and obstructions that were not readily apparent in two-dimensional drawings. Three-dimensional computer assisted design software tools now allow the designers and builders to "see" how the submarine arrangements and cable and pipe runs interact on the computer without building a physical mock-up. However, the *Virginia* and *Seawolf* programs used prototyping and small mockups to examine the technology implications and human interfaces of new systems. In the United States and UK, combat systems are typically prototyped ashore at a government or contractor facility to prove concepts and test the integration of various subsystems.

This chapter presents the lessons that are appropriate when setting the operational requirements for a new class of submarines. Operational requirements include not only performance metrics such as speed, diving depth, and signatures, but also the operational availability of the platform and how the platform will be operated. The operational requirements for a new platform are translated to performance specifications, which lead to technology choices to achieve the desired performance. Technology risks can be reduced when program managers know existing technologies and understand how operational requirements relate to available technologies.

Remember That the Submarine Is an Integration of Various Systems

The submarine is an integration of the pressure hull, a power and propulsion system, sensor and communication suites, and weapon systems. Operational requirements in one area will affect design considerations in the other areas. More-capable sensor systems may require additional power and a different propulsion system, which could affect the pressure hull design. The desire for greater weapons capability with more or newer weapons may also affect pressure hull dimensions.

It is challenging to find the right balance among the various system requirements especially when doing so for a submarine class that will be in the operational fleet for 30 years or more. Operational requirements and technologies change over time resulting in major modifications during a submarine's operational life.[1] When setting the requirements for different submarine systems, a program must understand the current and emerging technologies in those systems, how requirements might change in the future, and the trade-offs between costs and risks (the subject of the next lesson).

---

[1] The initial design of a new submarine will include margins for power, weight, and other metrics. The programs we studied maintained adequate design margins during the design and construction of the class. This practice should continue for future programs.

Understand the Current State of Technology to Control Program Risks

Program managers must understand the current state of technology in areas related to their programs. They also must understand how a platform's operational requirements affect technologies, risks, and costs. The desired operational performance will drive the characteristics of a platform and the technologies needed to achieve that performance. Program managers must be supported by a technical community (as mentioned in the previous chapter) that completely understands the technologies that are important to the program, where those technologies exist, and which technologies must be significantly advanced.

Additionally, it is important for program managers to understand how changes to operational requirements relate to the technology levels that are available. That is, if certain operational goals are beyond the state of current technology, what operational capabilities can be supported by existing technologies? This involves an understanding of trade-offs between operational requirements and technological risks (and costs). Again, this is where both operators and the technical community are important during the early stages of a program.

Relying on significant advances in technology may be necessary in some instances. During both the *Seawolf* program and the initial stages of the *Astute* program, the United States and UK felt that they needed to significantly expand their submarine capabilities to meet an increasing Soviet threat. And with the *Collins* program, Australia demonstrated its desire to have a platform with operational capabilities that exceeded those that were available in existing conventional submarines. When multiple new technologies are required, it is important for those involved with a new submarine program to recognize the risks and factor them into cost and schedule estimates.

Involve All Appropriate Organizations When Setting Operational Requirements

One shortfall of the *Astute* and *Collins* programs was the failure to involve all knowledgeable organizations in setting requirements. Both programs suffered when operators, maintainers, and builders did not have inputs during the early stages of the program. The program manager must be supported by adequate technical, operational, and management expertise in the government and private sector. This is especially important when setting requirements early in the program. Technical experts in laboratories and test centers can keep the program manager informed about existing and new technologies. Navy operators can provide insights into current submarine missions and capabilities, and private-sector companies that maintain submarines can provide information about how designs and operational requirements influence support costs. Experienced designers and builders can shed light on the difficulties and costs of achieving certain operational objectives. Moreover, these experienced designers and builders can help government engineers and acquisition experts draft contract specifications that achieve desired performance and safety outcomes in a manner clearly understood by all parties.

Involving various organizations is important throughout the life of the program. The program manager should have authority to make decisions based on various technical and operational resources. Also, involving all appropriate organizations helps develop knowledgeable, experienced managers for future programs.

Clearly State Operational Requirements

Operational requirements must be clearly stated as the desired performance of the submarine in various key areas. Key areas include speed, payload, and signatures, as well as other characteristics such as crew size and operational availability. These performance requirements must be backed by technical specifications, especially in the area of safety. Requirements should not be stated as point solutions but rather as objectives and thresholds, and those in the program must understand the cost and performance implications of meeting the threshold and the objective levels in the various key performance areas.

The United States, with its long history of submarine programs, appears to have learned this lesson. Issues rarely arise between the government and a prime contractor concerning desired operational requirements and the capabilities of the new submarine class.

In Australia, the operational requirements for the *Collins*, although mostly beyond the capabilities of conventional submarines available at the start of the program, were straightforward. Several key parameters were used to define the performance of the submarine.

The contract requirements for *Astute* were a mix of high-level performance attributes (e.g., speeds, signatures) and thousands of detailed requirements, technical specifications, and standards that were at times conflicting and difficult to interpret. The difficulties in clearly stating requirements for the *Astute* program were partly due to the shift of roles previously played by the Royal Navy and the MOD to the prime contractor, along with the desire to level the playing field between experienced and inexperienced competitors. In this environment, the government had no historical base on which to build.[2]

The government should state the desired performance of the platform but should avoid specifying how that performance should be achieved. The prime contractor should have the ability to decide how best to meet performance requirements. At times, there will be a benefit in designating a preferred provider or material. In those instances, the prime contractor should have the expertise to evaluate the requirement and suggest alternatives if appropriate. The *Collins* contract imposed both performance criteria and some detailed specifications on how the performance should be achieved. The contract was a mix of requirements and specific solutions; in some cases, the solutions could not meet the requirements. This became a problem with the combat system (e.g., specifying the use of the Ada programming language[3]), the propeller (e.g., specifying the use of Sonoston), and the periscopes (e.g., the use of the supplier of the *Oberon* periscopes).

---

[2]   The MOD has recognized the complexity of the initial contract requirements and is using performance-based specifications for the remainder of the first three submarines in the class. For the subsequent submarines in the class, a product build and test specification forms the basis of the contracts.

[3]   The *Collins* program was not alone in specifying the Ada programming language. The United States also specified Ada for its AN/BSY-2 combat system on the *Seawolf* submarine.

Requirements specification is a difficult balance of staying within known and approved standards and allowing innovation in the design, especially to reduce costs. The operational requirements should be supported by standards that relate to different functional systems. The prime contractor should be allowed to challenge standards and specifications if it can prove that the change will reduce cost or improve performance with the same or less risk.

Determine How to Test for the Achievement of Desired Operational Requirements

Stating an operational requirement is the first step in setting program goals. But that first step must be complemented by a plan to understand whether the platform meets the requirement. This typically involves test procedures—who will test, how the test will be conducted, and how success or failure will be measured. Although it is often difficult to plan tests early in a program, it is necessary to ensure all parties agree on the processes to measure how the performance of the platform meets operational capability objectives. Incremental testing of equipment before it becomes part of a system and before that system is inserted into the hull should be encouraged.[4]

With the *Collins* program, Australia had to learn for the first time how to test and accept a new submarine. Unfortunately, adequate testing procedures were not developed or enforced. For example, comprehensive tank testing of the hull design was not specified or accomplished, and the Hedemora engine configuration installed on *Collins* was not adequately tested before the submarines went to sea.

During the initial stages of the *Astute* program, the effort to reduce government overhead costs led to the deactivation or downsizing of the Royal Navy and MOD technical organizations that had overseen the testing and commissioning of all prior UK nuclear submarines. Without this knowledge and expertise, testing was largely ignored during

---

[4] An example of testing a major system before it is inserted into the submarine is Electric Boat's command and control system module off-hull assembly and test site (COATS).

the contract negotiations and early stages of the program. With the first-of-class, both parties struggled to identify and approve procedures for testing whether the vessel's performance met requirements.

CHAPTER FOUR

Lessons When Establishing an Acquisition and Contracting Environment

Establishing an open and fair acquisition and contract environment is an important aspect of any program. Poor decisions here will resonate throughout the life of the program. Issues include choosing the organizations involved in designing and building the new submarine, the type of contract, the specifics within the contract (including incentives), the decisionmaking process to employ when issues arise, and the payment schedule. The lessons often overlap but aim for a fair, interactive partnership among the program office, prime contractor, and subcontractors. Overall, the program should be a partnership between government and private-sector organizations. Both sides should work together toward the common goal of program success.

The acquisition and contract strategy can foster or hinder the desired interactions and relationships between the government and the private sector. In Australia, the relationship between the program office and ASC, the prime contractor, was strained during the conduct of the *Collins* program.[1] The differences grew out of many issues in the contracting environment and greatly affected the conduct of the program.

---

[1] The McIntosh and Prescott report stated ". . . the positions of the parties (the operational RAN, the procurement project office, the in-service support project team, the prime contractor, and the principal sub-contractors) are certainly far more antagonistic, defensive, uncooperative and at cross-purposes than should be the case in a project like this." See McIntosh and Prescott, 1999, p. 8. That report also includes the following observation by Lloyds Register: "In looking at ASC's conduct throughout the review period, there appears to be an underlying atmosphere of confrontation and contempt for their customer's wishes, with no visible recognition that their customer was and is unhappy and what could ASC do to rectify the matter" (p. 10).

In the UK, the shift in roles and responsibilities at the beginning of the *Astute* program resulted in a "hands off" approach by the MOD. This lack of oversight and integration with the prime contractor blinded the MOD to the problems in the program. The lessons we describe below will help to create the desired relationships and degree of mutual understanding among all parties.

Consider a Single Design/Build Contract for the First-of-Class

Typically, a submarine program involves an initial contract to design and possibly build one or more submarines and subsequent contracts to build the remaining submarines in the class.[2] The initial contract is the most important and sets the tone for the rest of the program. A common lesson across all five programs is that one prime contractor should design and build the first-of-class. A single design/build contract helps to integrate the two processes and reduces confusion and misinterpretations. Contracts for subsequent boats could be competitive but should be timed such that the design of the new submarine is largely fixed and the build process is well understood.

In the United States, the *Ohio* program had one organization, Electric Boat, design and build the submarines but contracted with different Electric Boat divisions to design and build the first-of-class (with the design provided to the builder by the U.S. Navy). Reconciling differences between the two contracts entailed schedule delays and cost growth. In the *Seawolf* program, the two shipbuilders each designed portions of the ship but competed to build the first-of-class. Again, there were significant problems with this approach. The *Virginia* program involves a single design/build prime contractor, Electric Boat, with Newport News serving as a major subcontractor to Electric Boat. This arrangement, plus other initiatives, has resulted in a largely successful program.

---

[2]   There are exceptions, of course. For example, the sole *Collins* contract was for the design of the submarine plus the build of all six in the class.

In the UK, the *Astute* program had a single firm design and build the submarine. The *Astute* program probably made the right decision in having a single prime contractor; the problem during its early stages stemmed from the decision to hold a competition when only one company had experience in designing and building submarines. The low winning bidder had never designed or built a submarine before, but agreed to further price reductions during negotiations prior to contract award. Further, the lack of integration between the design and the build teams, caused by the distance between the prime contractor design office and the shipyard—both in miles and relationships—contributed to many of the early problems. The original *Astute* contract also included building the first three submarines, not just the lead boat.

The Australian *Collins* program used a Swedish design (produced by Kockums) that was built by the ASC, an entity whose ownership included the government, Kockums, and other firms. Kockums was therefore involved in both the design and the build of the submarines. One main problem with the *Collins* program was the design organization's lack of appreciation for the demands of the concept of operations and the operating environment. Kockums was a successful designer of submarines for the Swedish Navy, but those submarines operated in a far different manner and in a different environment from what was planned for the *Collins* boats. The very different operating environments required different equipment and different procedures for operating the equipment. Therefore, it is important for the design organization to fully understand and appreciate the way the new submarines will operate and the impact of the operational environment on the design of the boats.

The choice of organizations and their role in a new program must reflect the status of the industrial base and the policy on potential future competition for design and build contracts. Currently, there is little or no chance for competition in the three countries. The UK and Australia have only one firm with experience in designing and building submarines. The United States has two nuclear submarine shipbuilders—Electric Boat and Newport News. Once competitors, these two firms have formed an effective partnership for the *Virginia* program. Given the direction of future defense budgets and the gaps between new program starts, it is unlikely that the governments could conduct, or could afford, meaningful competition for new submarine design and construction.

Even with a single contract for the initial design and the construction of the lead ship, the lead ship should be priced only when the detailed design is sufficiently complete for both the shipbuilder and the navy to have enough knowledge to estimate realistic cost.

Use a Contract Structure with Provisions to Handle Program Risks

The UK MOD used a fixed, maximum-price contract with the *Astute* prime contractor. Unfortunately, both the MOD and the prime contractor underestimated the substantial risks in having such a lengthy lapse of time between the *Astute* and its predecessor submarine program and in transferring responsibilities to the prime contractor. The two parties also overestimated the benefits of three-dimensional computer assisted design software and of the modular build process. The result: a program that could not achieve its original contract price and an environment in which (1) BAE Systems had no motivation to provide more than what it interpreted were its obligations in a contract with ill-defined specifications, and (2) the MOD was afraid to enforce ill-defined specifications for fear of being liable for contract changes that it could not pay for.

With the *Collins* program, although there were a number of technical risks with unpredictable outcomes, the Australian government used a fixed-price contract that greatly limited the flexibility that both parties needed when problems emerged. As with the *Astute*, the fixed-price contract for *Collins* led to an environment in which ASC had no motivation to provide more than what it interpreted were its obligations under a poorly defined contract. At the same time, the Commonwealth, fearful that it might be held liable for contract changes it could not afford, paid no more than the original contract price. The interactive and open environment necessary for a development program was negated by the *Collins* contract.

In the United States, the *Ohio* and *Seawolf* lead ship contracts were both fixed-price and incentive-type. But the risk sharing was substantially different from the *Los Angeles*–class early ship contracts. Both had escalation provisions that covered the effects of inflation up to the ceiling price and to the contract delivery date, without penalty. Both had substantially larger spreads from target cost to ceiling price than the early *Los Angeles*–class contracts. The *Virginia* program's lead ship risk provisions took a different approach. Rather than providing the detailed design drawings (developed under a separate contract) as government-furnished information to the construction shipyard, the *Virginia* program added cost-plus-incentive-fee construction line items for the lead ship to the original cost-plus design contract.

Fixed-price contracts are appropriate when there is little risk and uncertainty (e.g., when technologies are mature and when specifications are well defined) and when few changes to the design or build are anticipated. Although the government can try to place all risk on the contractor through use of a fixed-price contract, the government ultimately holds certain program risks. It is far better to structure a contract that holds the contractor responsible for risks under its control (labor and overhead rates, productivity, materiel costs, etc.) but holds the government responsible for risks beyond the contractor's control (inflation, changing requirements, changes in law, etc.). Otherwise, contractors will greatly increase their bid prices to accommodate risks that they cannot control. Appropriate cost-sharing provisions can be drafted to handle risks that neither party controls or that both parties have equal influence over (technology changes, acts of God, energy shortages, etc.).

Any contract, whether fixed-price or cost-plus, must have adequate incentives for the contractor to do better and to improve on the cost, schedule, and performance goals set by the government. The lesson here is that technical risks must be identified early, and much thought must be given to deciding, with industry, the appropriate form of the contract and the incentive and risk sharing clauses built into the contract. Getting this wrong can almost guarantee problems with the conduct of the program and the relationships between the government and the contractor.

Develop Realistic Cost and Schedule Estimates

Both the government and the private sector should develop cost and schedule estimates for designing and building the submarine. These should be realistic and based on the best knowledge and information available. Costs should be categorized at an appropriate level of detail to allow comparisons between the government and private-sector assumptions, methodologies, and estimates. Any discrepancies in cost estimates should be understood and discussed between the two parties. To support this requirement, the government should have a cost-estimating organization that collects, stores, and analyzes data on previous submarine design and build programs. This was one shortfall for the *Collins* program, since the Australian government did not have an independent cost-estimating process that fully understood the potential costs and risks.

It is important that both sides agree on the assumptions that underlie the cost and schedule estimates, including the impact that different risks could have on final costs. A program manager should avoid accepting cost and schedule estimates that are overly optimistic or that fail to address program risks. The contract cost for the UK's *Astute* program was based on unrealistic assumptions about the benefits of three-dimensional computer assisted design and manufacture software tools and of workload reductions resulting from a modular build process. The cost estimates also underestimated the difficulties in reconstituting a design and production capability after a substantial lapse in program starts and in transferring responsibilities to the private sector. The mantra "if it looks too good to be true, it probably is" applied to the *Astute* contract price in hindsight. The risks involved with several new technologies were underestimated by the U.S. *Seawolf* program, which resulted in the program's costs growing significantly.

The government should update cost and schedule estimates when new data and information are available. There is a tendency in programs to fix the anticipated delivery date of the first submarine at the time of the contract but not to adjust this date when problems emerge. Program managers should adjust future schedules when delays occur in a program. As we discuss in the next chapter, it is important that the build begin only after the submarine arrangements are complete and the majority of design drawings are produced. When delays occur in the design process, the start of construction and the planned delivery of the submarine to the government should be extended.

Decide on Government-Furnished Equipment

One important decision when establishing the acquisition strategy is which equipment will be bought and managed by the government and supplied to the builder as government-furnished equipment (GFE) or government-furnished information (GFI) and which equipment will be bought and managed by the contractor (contractor-furnished equipment, CFE). These decisions are based on many factors, including which party—the government or the private sector—holds risks and responsibilities in different areas, and which is better positioned to manage the subcontractors and the integration of the equipment into the submarine.

The GFE-versus-CFE decision for the United States and the UK is especially important for the nuclear reactor. The United States has always provided the reactor to the build contractor as GFE. For *Astute*, Rolls-Royce, the sole provider of UK nuclear steam raising plants (NSRPs) for submarines, was assigned as a subcontractor to the prime contractor (in essence making the NSRP CFE). Because safety risks reside with the MOD and Rolls-Royce has a contractual relationship with the MOD for in-service submarine support, the NSRP probably should have been GFE for the *Astute*.

Another important area when making GFE-versus-CFE decisions is the combat system. One issue that plagued the *Collins* program was the relationship, or lack of one, between the platform prime (ASC) and the combat system prime (Rockwell). The Commonwealth negotiated the contract with Rockwell but made ASC the prime contractor responsible for the successful delivery of the combat system, even though ASC played no role in choosing Rockwell and initially had no access to the classified specifications. If the combat system is to be a new design, it should probably be GFE, at least until delivery of the first ship in the class. This is especially the case for new combat system designs that push the technology of existing designs. The prime contractor or a third party can function as the system integrator, but the government may have the best leverage with the combat system designer.

One issue the government must weigh when making GFE-versus-CFE decisions is its ability to manage subcontractors to ensure that equipment is provided to the shipbuilder when specified in the contract. Construction schedules are tied to the planned delivery of various systems and pieces of equipment. Delays in those deliveries can lead to additional costs and delays. The problems between the contractors for the *Collins* combat system ultimately led to the Commonwealth being liable for delays with the delivery of the combat system.

Regardless of who provides the equipment and systems, it is necessary to sustain a viable vendor base. Suppliers go out of business if there is not an adequate demand for their products and services or if a new technology makes their product obsolete. Or suppliers may choose to stop a product line because of uncertainty in future demand. Both the government and the submarine design and build organizations must continually monitor the health of the vendor base and, when necessary, certify new suppliers. Supporting nuclear vendors is especially important for the United States and UK. The need to reconstitute suppliers can lead to cost and schedule growth. For example, one impact of the gap between *Vanguard* and *Astute* was the atrophy of the supplier base, resulting in the need to identify and certify many new suppliers.

Develop a Timely Decisionmaking Process to Manage Change

Changes invariably occur during any program. They may crop up in the desired performance of the platform; in the systems and equipment used to achieve performance; in the schedule of the project; or in the responsibilities of the various organizations involved in designing, building, and testing the platform. Management structures must be in place to deal in a timely manner with any contract changes that are proposed during the program. Changes may affect cost, schedule, or capability. It is important that the program office understand the impact of proposed changes and have a procedure in place to approve or reject them. This requires the involvement of the technical community, the cost estimation community, and the contractor. When funding is limited, changes that increase costs must be especially examined.

The government should have an adequate on-site presence at the design and build organizations. Minor issues that arise during the design and build can be adjudicated by that on-site presence. Major decisions on requirements changes, cost or schedule impacts, and equipment decisions should be made by the program management.

Establish an Agreed-Upon Tracking Mechanism and Payment Schedule

It is important to have an effective system to track progress and a payment schedule that is tied to clearly defined milestones and that reserves adequate funds to handle difficulties that occur later in the program. The progress tracking system must be properly used (e.g., accurate data reported on the correct project tasks) and have outputs that are helpful in managing the program. A tracking system by itself is not enough. There must also be an independent validation mechanism to confirm design and construction progress. The payment schedule in the contract should be tied to either a clearly defined and meaningful milestone plan or a well defined physical progression system. Adequate funds need to be reserved to handle difficulties that occur later in the program. Payment schedules should incentivize real progress and not encourage wrong behaviors.

The United States has effectively used earned value management (EVM) systems to monitor progress on its programs. During its first several years, the *Astute* program had no effective mechanisms to track progress on the design and build of the submarine. This made it impossible for the MOD, and even the prime contractor, to recognize problems that arose. One important change at Barrow, started by both the prime contractor and the MOD and assisted by Electric Boat, is the installation and use of an EVM system.

One shortcoming of the *Collins* program was paying the contractor the majority of the funds well before the project was complete. This led to having little or no funds available to handle problems as they arose later in the program. Also, the program office became aware of difficulties and problems too late in the process and was unable to make decisions that could have resulted in less costly corrections.

Include an Adequate Contingency Pool

Problems arise during new programs. All five cases we examined showed cost and schedule increases for the first-of-class. It is important that the contract include adequate contingency funds to cover unanticipated problems. The size of the contingency fund is related to the technical risks in the project—more risks require larger contingencies.

One criticism of the *Collins* program was that it lacked an adequate contingency to manage risks and changes. Where normally a complex project would have a contingency fund on the order of 10 to 15 percent, the *Collins* contract had only a 2.5 percent fund. This, along with having no agreed-upon processes for disbursement, undermined relations between the customer and the supplier and limited what the parties could do when problems arose.[3] The *Astute* program also had a very low fund for contingencies—about 5 percent of the contract value.

---

[3] Woolner, 2009.

CHAPTER FIVE

Lessons When Designing and Building the Submarine

Many lessons described in the previous chapters also apply to the design and construction phases of a new program. It is important to get all the appropriate organizations—operators, maintainers, and the technical community—involved throughout a program, to understand how operational requirements affect design and construction, and plan for the appropriate testing of the systems and platform to ensure requirements are met. Therefore, several lessons described below echo those described previously.

Involve Builders, Maintainers, Operators, and the Technical Community in the Design Process

One important lesson from the *Virginia* program is to use a design/build process during the design of a new submarine. This involves having the builders actively involved in the design process to ensure that what is designed can be built in an efficient manner. The design/build process should go further than merely involving builders in the design process. The design should also be informed by operators, key suppliers, maintainers, and the technical community. Therefore, it is important to think of the design team as a collaboration of submarine draftsmen and design engineers with inputs from those who must build to the design, operate the submarine, and maintain it. This collaboration should extend throughout the design process. However, throughout that process, it is important to keep in mind that the ultimate design and construction target is a submarine that is cost-effective in its post-delivery and ILS period of life. While maintenance ease is a desired trait, it must be balanced against long-term maintenance costs.

It is important not only to have the technical community in the design process but also to listen and react to the concerns it may raise. The degree to which existing technology is "pushed" in a new design will affect the risks to cost, schedule, and performance of the platform. The technical community must understand the state of technology and the degree to which a new design extends that technology.

The technical community consulted during a new design effort should extend beyond the in-country resources to include the technical assets of partner nations. In some areas, especially technical ones not encompassed in previous programs, other countries may have a deeper and better understanding of the technology and risks. For example, the Australian technical community may have knowledge of air-independent propulsion but very limited experience.

Complete the Majority of Design Drawings Before Start of Construction

An essential lesson for the build of a new submarine is to complete the majority of the design drawings before beginning construction. The *Collins, Seawolf,* and *Astute* programs all began construction well before the submarine arrangements were complete and with only a small percentage of the design drawings done. All three programs ultimately incurred additional costs for ripping out pipework, cabling, and equipment foundations that were installed too early, as well as for additional rework as the design matured. The *Virginia* program reversed this trend by having the arrangements finished and the majority of the design drawings complete before construction started. Although the complete drawing package does not have to be finished when construction starts, the drawings for a specific section of the submarine should be complete before construction starts on that section.

There is often a push to remain on schedule or to show progress to the government or the public. It is far better to delay construction to ensure that the design is largely complete rather than risk the costly rework and changes typically resulting from an immature design. Using three-dimensional product models facilitates the design/build process, but these models must be completed early so that material can be ordered and manufacturing data can be downloaded into numerically controlled machinery. Completing three-dimensional product models early in the process ensures that all pieces fit and minimizes expensive rework. A good rule of thumb is to have the electronic product model finished and 80 percent or more of the detailed design drawings complete when construction begins.

Ensure Sufficient Oversight at the Design and Build Organizations

At the beginning of the *Astute* program, MOD oversight at the Barrow shipyard was reduced greatly as part of the government's move to control spending. With no on-site presence, the MOD was blind to design and construction problems that cropped up in the early years of the program. The MOD has since increased its presence at Barrow to approximately 50 people (from a low of four) in hopes of having greater visibility and inputs into the program.

The program should have a strong presence at the shipyard to flag deviations from design, ensure compliance to quality and testing procedures, and keep the government aware of the challenges that the program faces. As mentioned in the previous chapter, the on-site government representatives should also have some decisionmaking capability in order to facilitate concessions and deviations that have only a minor impact on cost, schedule, or performance.

Specify and Manage Adequate Design Margins

A new submarine design must include adequate weight, stability, power, cooling, and bandwidth margins, all of which must be closely managed during the design, build, and operation phases. New ships and submarines typically start with what are believed to be adequate design margins. But often these are consumed during the design/build process or early in the platform's life. This is a problem that the *Collins* is experiencing and that all submarines typically experience to some extent. Without adequate margins, it may not be possible to modernize and upgrade equipment. New power and cooling plants may be needed, but they may exceed available weight margins. Existing systems may be downgraded or ship operations may be constrained if adequate margins are not available.

Develop an Integrated Master Plan for Design and Build

The lack of an effective integrated master plan blinded the MOD to the *Astute* program's schedule problems. A program should have an overall integrated schedule detailing the tasks, milestones, and products that are expected during the design and build of the submarine. Furthermore, a new program must not only develop an integrated master plan but must aggressively manage the program to ensure that it stays on the schedule suggested by the plan. The integrated master plan shows the order of tasks and events and the interrelationships among them. It can indicate the critical path for achieving the program schedule and the impact on the schedule of delays in any task. A key decision is the level of detail to include in the plan. Although more detail can provide greater insights, it takes more effort to create and manage.

An integrated master plan is a first step in understanding the status of a program. The development of a system to monitor progress, the subject of the next lesson, is a necessary second step.

Track Progress During the Design/Build Process[1]

The previous chapter discussed the need to develop a tracking system so that the government can understand the status of a program and foresee problems before they actually occur. Here, we stress that such

---

[1] See Arena et al., 2005, for a specific description of EVM.

a system must be in use during the design and build and utilized correctly.

During the first several years of the *Astute* program, there was no effective system to monitor the progress of the design and build. Ultimately, an EVM system was put in place. However, EVM represents a cultural change for the shipyard, and workers still find it difficult at times to allocate the proper data to the right project or task. An accurate cost accounting system is a necessary prerequisite for a meaningful EVM system.[2]

It should be noted that EVM has a number of limitations. It provides few, or even incorrect, insights if the proper data are not collected and reported correctly. EVM also lacks flow and value-generation concepts.[3] Because building in a proscribed sequence is so critical in submarine programs, EVM must be used with care to avoid introducing bad behaviors.

Whether EVM or another progress monitoring metric is used, it is important to have an effective system to track progress and predict cost and schedule status. It is also important to have an independent validation procedure to confirm the progress suggested by the tracking system. The Supervisor of Shipbuilding at the shipbuilders' facilities in the United States and the MOD oversight group at the Barrow shipyard provide this independent confirmation process.

---

[2] *Earned value metrics* compare the budgeted cost of work performed (BCWP) with the budgeted cost of work scheduled (BCWS) at a given point in time. When the BCWP value is less than that of the BCWS, the project is considered behind schedule. If the BCWP value exceeds the BSWS value, the project is considered ahead of schedule. The *schedule performance index* is equal to BCWP divided by BSWS. The *cost performance index* is the BCWP divided by the actual cost of work performed. Indices less than 1 indicate the project is behind schedule and over budget.

[3] *Flow* refers to how resources and activities are sequentially related. *Value-generation work* is work performed in one time period that will allow future work to begin.

Design for Removal and Replacement of Equipment

The operational life of a submarine is typically longer than the life of some of the technologies it employs. This is especially true for command, control, communications, computing, and intelligence (C4I) equipment. Adequate access paths and large equipment removal hatches were included in the *Ohio, Seawolf,* and *Virginia* designs, facilitating the removal and replacement of equipment that requires repair or has become obsolete. However, the *Collins* and *Astute* designs did not include adequate routes for equipment removal and replacement so the logistics support of those platforms will likely be much more difficult (a topic for the next chapter). For those submarines, large hull cuts may be required to remove and replace equipment that becomes obsolete or needs repair.

The design of the submarine should anticipate the need to remove and replace large pieces of equipment and include access paths and hatches to do so. For C4I equipment, modularity and interoperability should be incorporated into the design.[4] Data and information architectures should be developed that allow installation of electronic equipment as late in the build process as possible to take advantage of the rapid changes in information technology.

Conduct a Thorough and Adequate Testing Program

The previous chapter discussed how a new program must specify not only desired operational requirements but also test procedures to ensure that those requirements have been met. The test procedures should be developed during the design/build portion of the program. Testing should involve the design and build organization(s), as well as the technical community and the navy.

---

[4]   See Schank et al., 2009.

CHAPTER SIX

Lessons for Integrated Logistics Support

Integrated logistics support begins more than a decade after a submarine is initially designed. But despite that time gap, ILS needs to be incorporated in all early planning for a submarine; it must inform the submarine's design and construction and help structure the facilities, contracts, and procedures that will be required to keep the vessel operationally available.

Typically, a submarine's operating and support costs over the course of its service life are much greater than its initial acquisition cost. But design/build programs often unwisely focus on reducing the platform's unit procurement cost rather than its whole-life cost. It is difficult to convince senior decisionmakers to spend more money in the short term to save greater amounts in the long term. Therefore, persuasive arguments are necessary to ensure that the costs of integrated logistics support are considered during the design/build process.

Australia's current problems with the operational availability of the *Collins* class largely resulted from the lack of developing a thorough ILS plan during design and construction. Although ILS planning was included in the original contract with ASC, funding for developing the plan was systematically reduced to address other issues during design and construction. A strategic view of ILS early in the program was particularly needed because the RAN was thrust into the unfamiliar role of a parent navy with the *Collins*. The original plan of "business as usual" failed to consider the unique requirements of maintaining the submarines and training their crews.

Similarly, in the UK, the original *Astute* contract included several boat years of contractor logistics support for the first three submarines in the class. The intent was to have the prime contractor design the submarine in a way that would reduce the whole-life cost of the platform. However, the prime contractor lacked the expertise to fully understand in-service support requirements, risks, and costs. When the contract was renegotiated, the contractor logistics support provisions were deleted with an agreement that the ILS costs would be established at a later date. Because of the cost and schedule problems with the *Astute*, the prime contractor focused on reducing the procurement cost of the platforms rather than reducing the whole-life costs of the submarines.

In this chapter, we look at important lessons when establishing an ILS program. These include developing an ILS strategic plan during the design phases, taking an integrated government view of submarine support versus an isolated programmatic view, establishing a planning yard function to track maintenance and reliability and to project future maintenance needs, planning for crew training, and establishing and protecting the funding necessary to develop a comprehensive ILS plan.

Establish a Strategic Plan for ILS During the Design Phase

A strategic plan for ILS must be started early in the program, preferably during the design phase. As mentioned in the design and build lessons, personnel from the organizations responsible for maintaining the submarine should be involved in the design process to ensure that what is ultimately built can be efficiently and effectively supported. Funding should be established to develop the ILS plan and should be protected during program execution.

A strategic ILS plan is predicated on the following tenets:

- Maximize equipment commonality by standardizing parts.
- Support operational availability by testing the reliability of equipment.
- Ensure maintenance ease and accessibility by taking into consideration the long-term costs that may be incurred.

- Take into account plans for technology and capability development over the operational life of the class.

An ILS plan depends upon establishing a concept for operating and maintaining the submarine. The concept of operations must recognize that the submarine will require time for preventive and corrective maintenance and for equipment modernizations, and the plan should account for periodic cycles of training, operations, and maintenance that hold throughout the life of the submarine. It goes without saying that developing this concept of operations and maintenance must involve both operators and maintainers.

In order to develop a maintenance plan, the reliability and maintainability of the equipment and the need for corrosion control of the hull must be well understood. This involves frequent interactions with the design authorities and the original equipment manufacturers (OEMs) to obtain data. It also involves a thorough understanding, informed by a robust database, of the reliability and maintainability of new and existing equipment used in the new platform. Equipment should be thoroughly tested under conditions and missions that the submarine is expected to encounter throughout its service life. Maximizing the use of standard or common systems, equipment, and parts whenever possible in the design can provide valuable insights into reliability and maintenance.

The strategic plan for ILS should identify when maintenance, modernization, and training will be performed, where the activities will take place, and which organizations will be involved. It should provide clear guidance on how maintenance activities should be conducted. Equipment reliability and the need for corrosion control will factor into when maintenance should be performed. Some maintenance will be the responsibility of the crew at the operating base; higher-level maintenance and modernization will be the responsibility of government or private-sector organizations and will be accomplished either at the operating base or at a shipyard. As discussed above, the end result should be a thorough plan for maintenance and modernization throughout the life of the submarine.

Finally, the ILS strategic plan must include provisions to modernize equipment during the submarine's operational life. It is inevitable that some equipment, especially electronics, will require updates. Modernizations may involve the higher-level maintenance organization but will more likely involve the OEMs. Electronic equipment may require time-phased upgrades involving both hardware and software. Setting periodic hardware and software upgrades will establish a drumbeat of modernizations throughout the program.

Consider ILS from a Navy-Wide Versus a Program Perspective

ILS must be considered at the force level, not at the specific program level. There will be demands on maintenance and training resources from older submarines still in the fleet (i.e., those being replaced by the new program), as well as surface ships. Those in charge of the new program must recognize these other demands and plan accordingly. This is especially important for limited maintenance facilities such as dry docks that are used across several classes of ships or submarines.

With the *Astute* program, the UK MOD included several years of contractor logistics support in the initial contract. The hope was the contractor would design the submarine to reduce the cost of in-service support. Unfortunately, early problems with the program caused the prime contractor to focus on reducing procurement costs and to ignore the impact on logistics costs. When the contract was renegotiated, the contractor logistics support provisions were removed because the MOD realized that the *Astute* submarines should be supported using the same process as the previous classes of submarines.

Establish a Planning Yard Function and Develop a Maintenance and Reliability Database

The original plans for ILS are likely to be modified as experience is gained on the reliability and maintainability of the equipment. Some equipment may require more maintenance than originally thought, while other equipment may prove to be more reliable or easier to maintain. Establishing a planning yard function that tracks maintenance and establishes future workloads is important to ensure that the right maintenance is done at the right times. This planning yard function can be performed by a government organization or by a private-sector firm. One function of the planning yard is to monitor and update data on the maintenance history of the new submarine. Another function is to stay in constant contact with the design authorities and OEMs to understand any changes in the platform or in equipment maintenance requirements and procedures.

Plan for Crew Training and Transition to the Fleet

The ILS plan must also include the "when, where, and who" for training activities. As with maintenance, some training will occur at the operating base while other training will be accomplished at centralized facilities. It may be done either by the navy or by a government or private-sector firm. When establishing the training plan, it is important to consider the transition of crews and personnel from an existing platform to the new submarine class. Also, a crew should be assigned to a submarine during construction so that the personnel can become familiar with the submarine and its systems and provide feedback during the build process. When a new submarine is delivered to the fleet, its crew should have been with the boat long enough to be familiar with all the vessel's operating procedures. Part of the training plan should identify when and how simulators or other training devices will be used to accomplish the training.

Maintain Adequate Funding to Develop and Execute the ILS Plan

Most important, there must be sufficient funds to develop and execute the strategic ILS plan. These funds should be "protected" during the design and build of the platforms. This was not the case with the *Collins*, whose original funding was systematically reduced to address other emerging problems during the design and build of the boats. As a result, the program lacked a thorough strategic ILS plan when the submarines entered service.

CHAPTER SEVEN
Summarizing the Lessons

We found numerous lessons in the five programs that the three countries pursued, and in the last several chapters we attempted to identify the major ones. But identifying lessons is merely the first step. Equally, if not more important for future submarine programs in the United States, the UK, and Australia is that policymakers and program managers learn these lessons and not forget them.

The important issue is recognizing the context in which decisions were made and the potential outcomes of those decisions. Each program was conducted in a different threat and budget environment, and some faced significant changes in government policies and in the health of the submarine industrial base. Often, the problems a program experienced resulted from underestimating, or ignoring, the impact of significant changes from previous programs.

Program costs are typically the primary focus of program managers and those who judge the success of a program. Therefore, the majority of the lessons are aimed at controlling program costs. We have separated the lessons across the five programs into those that apply at the top, strategic level of total submarine force management and those that apply at the specific program management level. The programmatic lessons are further distinguished by when they occur within a program—during the early stages when setting operational requirements and establishing an acquisition and contracting environment; during the middle stages of designing and building the submarine; and during the later stages when supporting the in-service submarines.

Top-Level, Strategic Lessons

The top-level, strategic lessons go beyond a single program or a single point in time and stress the long-term view of a country's overall submarine force and the industrial base that supports that force.

The government must be an intelligent, informed partner in the submarine enterprise. Being an intelligent partner requires a thorough understanding of the technologies applicable to a new program and the risks involved in extending existing technologies or relying on new technologies. This implies that all appropriate organizations, including operators, maintainers, the technical community, and the private-sector designers and builders, be involved with a new program from its beginnings. Being an intelligent partner also requires understanding the roles and responsibilities of the government and the private sector and developing a plan to mitigate and manage the risks that may arise during program execution. Finally, being an intelligent partner requires having knowledgeable and experienced people involved with the program. Future program managers must be identified early in their careers and provided opportunities to learn and grow the expertise needed to manage future programs.

Being an informed customer means that there is sufficient government involvement in and oversight of the design and construction processes to gauge the status of a program, identify where problems may arise, and work with the prime contractor and major subcontractors to handle problems in a way that avoids or minimizes cost and schedule impacts.

The government must take a long-term view of the submarine force and the industrial base. It must decide whether it is more cost-effective to sustain resources during any gaps between new submarine design and construction programs or whether resources should be allowed to atrophy and be rebuilt when needed in the future. There are advantages and disadvantages to each approach and the cost and schedule implications should be carefully thought out.

A submarine program needs the support of the legislative branch and the public. It is important that new programs be open and transparent to all parties, including government, the military, legislative bodies, and the public.

Programmatic Lessons

Setting Operational Requirements

A new submarine program must seek to reduce risks to the maximum extent possible. Reducing risks involves decisions made early in a program on the desired operational performance of the new submarine. Because a submarine is an integration of multiple systems, requirements in one area, such as speed, C4I capability, or weapons payload, can affect other areas and influence the overall design of the submarine. Therefore, it is important to understand the current state of the technologies that are appropriate to the program and be familiar with those new technologies that are needed to achieve the desired operational capability. Pushing technologies too far in too many areas can increase costs and delay schedules. To understand technologies and how they relate to desired operational capabilities all appropriate organizations must be involved in setting the desired performance goals for a program and have inputs into how those performance goals could be achieved. This includes operators of existing submarines, technology centers that monitor and understand the status of current technologies, designers who incorporate technologies to achieve performance goals, and builders who construct the submarines.

Operational requirements must be clearly stated as the desired performance of the submarine in various key areas, including speed, payload, signatures, crew size, and operational capability. Operational requirements should not be stated as point values but rather as ranges that recognize that there may be a desired objective value for a performance parameter but that a lower, threshold value may provide an acceptable solution at a lower cost. It is not sufficient merely to state the desired operational requirements. Procedures to test for the achievement of the desired operational requirements must also be determined early in the program.

Establishing the Acquisition and Contracting Environment

Establishing an open and fair acquisition and contracting environment will reduce problems that can occur during a new program. The acquisition and contracting strategy should aim for a partnership between the government (the buyer) and the private sector (the seller). Decisions made in this area can foster or hinder the desired interactions and relationships between the two sides.

A single contract that covers the design of the new submarine and the build of the first-of-class seems appropriate in most cases. Although acquisition and contracting decisions will largely depend on the status of the industrial base that designs and builds the submarines, having one organization responsible for the design and build can reduce communication and interface problems. The UK and Australia both have a single firm with experience in leading new submarine programs. The United States has two firms that design and build submarines, but those firms have formed a solid partnership that may preclude future competitions. Contracting for just the first submarine in a class will allow the design to be fixed and build problems to be resolved before contracts are let for subsequent boats in the class.

New programs must have an appropriate contract structure. Although fixed-price contracts can reduce risks of cost growth to the government, they are most appropriate when there is little program risk and uncertainty and when few changes are anticipated. With the risks and uncertainty of a new program, especially one that differs in some way from previous programs, a cost type of contract is probably most appropriate. Whatever type of contract is used, both the government and the private sector should develop realistic cost and schedule estimates. Any differences in the cost estimates of the government and the private sector should be understood and discussed between the two parties with the ultimate goal of agreeing on the estimates and schedules.

One important decision when establishing the acquisition and contracting environment is which equipment will be bought and managed by the government (as GFE) and which equipment will be bought and managed by the contractor (as CFE). This choice will largely depend on who holds various risks and responsibilities and who is better positioned to manage the subcontractors. Typically, the nuclear propulsion system and the combat system are candidates for GFE, while other major equipment and systems might be better managed by the prime contractor.

An open and fair acquisition and contracting environment should include a timely decisionmaking process to manage changes that occur during the program. Changes in desired performance, in systems and equipment, in schedules, in construction methods, or in responsibilities invariably occur during a new program. It is important that any such changes be adjudicated fairly and quickly to avoid schedule delays and cost growth. A mechanism must also be identified to track progress and provide progress payments to the contractor. The payment schedule should be tied to a clearly defined and meaningful milestone plan or a well-defined physical completion system. Finally, the contract should include an adequate contingency pool to handle any problems that arise in the program.

Designing and Building the Submarine

Operators, builders, and maintainers must be involved in the design process. Many of the lessons described above are also applicable when designing and building the submarine. We have already stressed the need to involve all appropriate government and private-sector organizations during the early stages of a program. The "design/build/maintain" environment during the design phases can help ensure that the submarine is built and supported in a cost-effective manner.

The majority of the detailed design drawings should be complete before construction begins. Starting construction before arrangements are complete and the design is largely fixed typically results in future rework to remove, modify, and replace structural and equipment components. Although there may be a reluctance to extend schedules set early in a program, it is often more cost-effective to delay the start of construction when design is still in flux. A rule of thumb that has emerged from previous programs is to have the arrangements fixed and the design drawings 80 percent complete.

The vessel's weight, stability, power, cooling, bandwidth, and other design margins should be closely managed. The design, build, and operations of the new submarine may begin to consume the initial design margins. Any change to the design margins should be thoroughly examined. Without adequate margins, it may be difficult or impossible to modernize the equipment or adjust the missions over the operational life of a submarine class.

The government must have sufficient oversight of the design/build process. The government needs to understand the status of a program, where problems may arise, and how best to resolve those problems. A strong and interactive government presence at the shipyards can help address deviations from designs, assure compliance to quality and test procedures, and keep the government aware of the challenges the program faces.

The government and the prime contractor, working together, should develop an integrated master plan to design, build, and test the submarine. The master plan should have an overall schedule linking tasks, milestones, and products during the conduct of the program. The government oversight resources should track the progress of the program to see if there is any deviation from the plan. Earned value management is one tracking tool used by various programs. But regardless of the specific tool or system used, it is important that the data be entered correctly and that the system provide forward looking performance measures.

The submarine should be designed for the removal and replacement of equipment. Modern submarines have operational lives of 30 years or more. During this time, equipment will require repair, removal, and modernization. The design of the submarine should anticipate the need to remove and replace various equipment and include adequate access and removal paths and hatches. Electronic systems for command, control, communications, and computing typically require frequent upgrades. The design process should evaluate different communications architectures and use standardized racks and equipment to the extent possible.

The government and prime contractor must conduct a thorough and adequate test program. The early stages of a new program should specify how the government and prime contractor should test subsystems, systems, and the new submarine to ensure that safety standards are met and performance goals are achieved. To the extent possible, systems should be tested before they are inserted into the hull. Testing should proceed during the build process and culminate in the various trials conducted by the shipbuilder and the government.

Providing Integrated Logistics Support

The government and prime contractor should establish a strategic plan for integrated logistics support during the design phase of a new program. Programs typically face pressures to reduce the construction costs of the submarines. Although logistics support costs occur more than a decade after the initial design of a new submarine, they represent the largest portion of a platform's life-cycle costs. Issues surrounding how the new submarine will be supported once it enters service must be addressed early and throughout the design/build process. The goal of a new program should be to reduce the total ownership cost of the fleet of new submarines, not the initial acquisition cost of a submarine. Therefore, the strategic ILS plan should include when, where, and how to maintain, repair, and modernize the platform. It should provide similar specifics on crew training and management. The plan should be based around a concept of operations for the submarine that details operational, training, maintenance, and modernization periods over the life of the submarine.

ILS should be considered from a navy-wide perspective rather than an individual program perspective. Other submarines and surface ships in the navy will also have strategic ILS plans. The various demands on logistics support resources should be coordinated to provide the required support to the fleet in the most cost-effective manner.

A planning yard function and a reliability and maintainability database should be included in the ILS plan. A planning yard can help track maintenance through the database and establish future support workloads and procedures. The planning yard function can be provided by either government or private-sector organizations.

There should be a plan for crew training and transitioning the submarine to the fleet. Training could be provided by a government or a private-sector organization depending on costs and capabilities.

The government must ensure that the program maintains adequate funding to develop and execute the ILS plan and to establish the support infrastructure. Budget pressures often threaten program funding and, since support costs occur much later, ILS planning is often one item where funding can be reduced. The program manager must remember that in-service support costs greatly exceed construction costs and that an effective ILS plan is necessary to reduce total ownership costs.

The implications of these lessons may vary for each country and for each future submarine program. The United States, due to its long, continuous history of designing and building new classes of submarines, appears to have learned the lessons from past programs and adapted them for new programs. The important, overall lesson for the United States is to not forget those lessons. The UK also has a long history of new submarine programs. The important lesson for the UK is always to understand the risks associated with major changes from previous programs and to plan for those risks. Finally, although Australia had a long history of submarine operations, it faced the responsibilities of designing, building, and supporting a new submarine for the first time with the *Collins* program. As with any first-time endeavor of a complex task, the *Collins* program experienced problems. The important lesson for Australia is not only to learn from the decisions and outcomes of the *Collins* program but also to draw from the experiences of the United States and the United Kingdom.

Bibliography

Arena, Mark V., John Birkler, John F. Schank, Jessie Riposo, and Clifford A. Grammich, *Monitoring the Progress of Shipbuilding Programmes: How Can the Defence Procurement Agency More Accurately Monitor Progress?* Santa Monica, Calif.: RAND Corporation, MG-235-MOD, 2005. As of June 27, 2011: http://www.rand.org/pubs/monographs/MG235.html

McIntosh, Malcolm K., and John B. Prescott, *Report to the Minister for Defence on the Collins Class Submarine and Related Matters*, Canberra, Australia: CanPrint Communications Pty Ltd., June 1999. As of June 27, 2011: http://www.defence.gov.au/minister/1999/collins.html

Schank, John F., Jessie Riposo, John Birkler, and James Chiesa, *The United Kingdom's Nuclear Submarine Industrial Base, Volume 1: Sustaining Design and Production Resources*, Santa Monica, Calif.: RAND Corporation, MG-326/1-MOD, 2005a. As of June 27, 2011: http://www.rand.org/pubs/monographs/MG326z1.html

Schank, John F., Cynthia R. Cook, Robert Murphy, James Chiesa, Hans Pung, and John Birkler, *The United Kingdom's Nuclear Submarine Industrial Base, Volume 2: Ministry of Defence Roles and Required Technical Resources*, Santa Monica, Calif.: RAND Corporation, MG-326/2-MOD, 2005b. As of June 27, 2011: http://www.rand.org/pubs/monographs/MG326z2.html

Schank, John F., Mark V. Arena, Paul DeLuca, Jessie Riposo, Kimberly Curry Hall, Todd Weeks, and James Chiesa, *Sustaining U.S. Nuclear Submarine Design Capabilities*, Santa Monica, Calif.: RAND Corporation, MG-608-NAVY, 2007. As of June 27, 2011: http://www.rand.org/pubs/monographs/MG608.html

Schank, John F., Christopher G. Pernin, Mark V. Arena, Carter C. Price, and Susan K. Woodward, *Controlling the Cost of C4I Upgrades on Naval Ships*, Santa Monica, Calif.: RAND Corporation, MG-907-NAVY, 2009. As of June 27, 2011: http://www.rand.org/pubs/monographs/MG907.html

Schank, John F., Cesse Ip, Frank W. Lacroix, Robert E. Murphy, Mark V. Arena, Kristy N. Kamarck, and Gordon T. Lee, *Learning from Experience, Volume II: Lessons from the U.S. Navy's Ohio, Seawolf, and Virginia Submarine Programs*, Santa Monica, Calif.: RAND Corporation, MG-1128/2-NAVY, 2011a. As of September 15, 2011:
http://www.rand.org/pubs/monographs/MG1128z2.html

Schank, John F., Frank W. Lacroix, Robert E. Murphy, Cesse Ip, Mark V. Arena, and Gordon T. Lee, *Learning from Experience, Volume III: Lessons from the United Kingdom's Astute Submarine Program*, Santa Monica, Calif.: RAND Corporation, MG-1128/3-NAVY, 2011b. As of September 15, 2011:
http://www.rand.org/pubs/monographs/MG1128z3.html

Schank, John F., Cesse Ip, Kristy N. Kamarck, Robert E. Murphy, Mark V. Arena, Frank W. Lacroix, and Gordon T. Lee, *Learning from Experience, Volume IV: Lessons from Australia's Collins Submarine Program*, Santa Monica, Calif.: RAND Corporation, MG-1128/4-NAVY, 2011c. As of September 15, 2011:
http://www.rand.org/pubs/monographs/MG1128z4.html

Woolner, Derek, "Taking the Past to the Future: The Collins Submarine Project and Sea 1000," *Security Challenges,* Vol. 5, No. 3, Spring 2009, pp. 47–71. As of June 27, 2011:
http://www.securitychallenges.org.au/ArticlePDFs/vol5no3Woolner.pdf

Yule, Peter, and Derek Woolner, *The Collins Class Submarine Story: Steel, Spies and Spin*, Cambridge, UK: Cambridge University Press, 2008.